# YOUR KNOWLEDGE HAS VALUE

- We will publish your bachelor's and
  master's thesis, essays and papers

- Your own eBook and book -
  sold worldwide in all relevant shops

- Earn money with each sale

Upload your text at www.GRIN.com
and publish for free

**Bibliographic information published by the German National Library:**

The German National Library lists this publication in the National Bibliography; detailed bibliographic data are available on the Internet at http://dnb.dnb.de .



**Imprint:**

Copyright © 2017 GRIN Verlag
Print and binding: Books on Demand GmbH, Norderstedt Germany
ISBN: 9783668965997

**This book at GRIN:**

https://www.grin.com/document/464858

Firdos Vani


# Integrated Nutrient Management in Underground Vegetable Crops

## On Integrated Nutrient Management


GRIN Verlag

## REVIEW ARTICLE

## Integrated nutrient management in underground vegetable crops

## F.B VANI

## Ph.D Scholar (Horticulture)


## Abstract

Vegetable comprises large number of plants, consumed as leaf, fruits, flowers, stem, roots etc. They are rich in nutrients like carbohydrates, proteins, fats, minerals and vitamins. They are mostly cultivated around the year throughout the country. India is the second largest producer of vegetables next to China in the world. It is cultivated in an area of 9575 ('000' ha) with production of 166608 ('000' MT) with the productivity of 17.40 MT/ha (NHB, 2016).

Now a days, modern agriculture depend heavily on use of chemical fertilizers for boosting crop yield. However, indiscriminate use of fertilizers has an adverse effect on long term soil health and environment which has global attention.

The realistic solution is Integrated Nutrient Management system are the combined application of chemical fertilizers, alongwith organic manure, green manure, bio-fertilizer and other organic recyclable materials for crop production.

## Introduction

Vegetable comprises large number of plants, consumed as leaf, fruits, flowers, stem, roots etc. They are rich in nutrients like carbohydrates, proteins, fats, minerals and vitamins. They are mostly cultivated around the year throughout the country. India is the second largest producer of vegetables next to China in the world. It is cultivated in an area of 9575 ('000' ha) with production of 166608 ('000' MT) with the productivity of 17.40 MT/ha (NHB, 2016).

Vegetable growing is the most remunerative enterprise as it is adopted on small and marginal holding with high production in short duration. Being a source of farm income, it creates impact on the agricultural development and economy of the country. Vegetables are cheaper source of minerals, vitamins and fiber with high calorific values. There is an increasing demand of vegetables both for domestic as well as for export, which can earn valuable foreign exchange for country.

### What is underground Vegetables?

> Any of various fleshy edible underground root, tubers or bulb is a underground vegetable.
> Vegetables play an important role in human nutrition. Underground Vegetables are defined as those annual or perennial plants where the economically useful parts like roots ,tubers etc are developed between the surface and subsurface layers of the soil which are commonly used for culinary purposes. And also called protective food as their consumption can prevent several diseases and are cheaper source of minerals, vitamins and fiber with high calorific values.
> Underground vegetables are modification of various plant parts like stem, root etc.

| Common name | Botanical name | Family | Chromosome number (2n) |
| --- | --- | --- | --- |
| Onion | *Allium cepa* L. | Alliaceae | 16 |
| Multiplier onion | *A. cepa* var. *aggregatum* L. | Alliaceae | 16 |
| Garlic | *A. sativum* L. | Alliaceae | 16 |
| Leek | *A. ampeloprasum* L. var. *Porrum* (L.) | Alliaceae | 32(4x) |
| Japanese onion | *A. fistulosum* | Alliaceae | 16 |
| Tree onion | *A. cepa* L. var. *viviparaum* | Alliaceae | 16 |
| Shallot | *A. cepa* var. *ascalonicum* L. | Alliaceae | 16 |
| Chive | *A.schenoprasum* L. | Alliaceae | 16, 24,32 |
| Chinese chive | *A. tuberosum* | Alliaceae | 32(4x) |
| Kurrat | *A. kurrat* | Alliaceae | 32(4x) |

- **List of underground vegetable crops  Bulb crops**

> ➢ **Root vegetable crops:**

| Common name | Botanical name | Family | Chromosome number(2n) |
| --- | --- | --- | --- |
| Carrot | *Daucus carota* | Apiacea | 18 |
| Radish | *Raphanus sativus* | Cruciferae | 18 |
| Beetroot | *Beta vulgaris* | Chenopodiaceae | 18 |
| Turnip | *Brassica rapa* | Cruciferae | 20 |

> ➢ **Tuber vegetable crops**

| Common name | Botanical name | Family | Chromosome number(2n) |
|---|---|---|---|
| Potato | *Solanum tuberosum* | Solanaceae | 48 |
| Sweet potato | *Ipomoea batatas* | Convolvulacae | 60 |
| Tapioca | *Manihot esculenta* | Euphorbiaceae | 36 |
| Elephant foot yam | *Amorphophallus campanulatus* | Araceae | 26 |
| Taro | *Colocasia esculenta* | Araceae | 28 |
| Tannia | *Xanthosoma sagittifolium* L. Schott | Araceae | 26 |
| Giant Taro | *Alocasia indica* (Roxb.) | Araceae | 28 |
| Yams | *Dioscorea spp.* | Dioscoreaceae | 40 |
| Greater yam | *Dioscorea alata* Linn | Dioscoreaceae | 40 |
| Lesser yam | *Dioscorea esculenta* Burk | Dioscoreaceae | 40 |
| White yam | *Dioscorea rotundata* Poir | Dioscoreaceae | 40 |
| Jerusalem Artichokes | *Helianthus tuberosus* | Compositae | 49 -53 |
| Coleus (Chinese potato) | *Solenostemon rotundifolius* (Poir) | Lamiaceae | 48 |
| Purple Arrowroot | *Canna indica* L. | Cannaceae | 18 |
| West Indian Arrowroot | *Maranta arundinacea* L. | Marantaceae | 18 |

- ➢ **Spices underground vegetable crops**
- ➢

| Common name | Botanical name | Family | Chromosome number(2n) |
|---|---|---|---|
| Turmeric | *Curcuma longa* | Zingiberaceae | 22 |
| Ginger | *Zingiber officinale* | Zingiberaceae | 22 |

**Importance of the underground vegetables**

- ❖ **Onion**
  - Onion bulb is a rich source of minerals like phosphorus, calcium and carbohydrates.
  - The antifungal properties of onion is due to presence of "catechol".
  - Onion is known to benefit in the prevention and treatment of atherosclerosis and coronary heart disease.

❖ **Garlic**
- It reduces the cholesterol in blood.
- It is rich in proteins, minerals like phosphorus, calcium, magnesium and carbohydrates.
- It also contains fat, vitamin C and sulphur.
- It is already being used in several food preparations, notably in chutneys, pickles, curry powders, curried vegetables, meat preparations, tomato ketchup.

❖ **Radish**
- Radish is good source of vitamin C.
- It helps to have sound bowel movement thus helping to prevent constipation.
- It is good for individuals with diabetes as radish helps to keep in check.
- It is good vegetable to have in winter as its helps to prevent from cold and cough.
- It also helps in the regulation of blood pressure.

❖ **Carrot**
- A sweet preparation Called Gajar halwa is Very famous dish.
- A special type of beverage knows as Kanji is prepared from black carrot which is used as appetizer.
- Carrot juice is used for the colouring of other foods.
- Rich in carotene, iron, thiamine, riboflavin, ascorbic acid, niacin, and sugar content

❖ **Beetroot**
- After sugarcane it is the second most important crop for the preparation of sugar.
- Beetroot is the rich source of protein, carbohydrate, Ca ,P, and Vitamin C.

❖ **Turnip**
- Turnip is mainly used as cooked vegetable, salad and pickle.
- In India leaves are used as green fodder for feeding for cattle which is rich in vitamins A and C, Calcium and iron.

❖ **Potato**
- The multifarious preparations are relished equally by the rich and the poor.
- Potato is a wholesome food.
- Potato is the protective food. It contains starch, sugar, cellulose, crude fiber, pectic substances, Protein, amino acids, organic auds, lipids, vitamin c, enzymes, minerals (P, Ca, Mg, K, Fe, S,) etc.
- Potato is the biggest source of carbohydrates, which comprises largely of starch. The calorically starch is most important nutritional component.

❖ **Sweet potato**
- Tubers are used for human consumption.
- Starch and alcohol are prepared.
- Young tender leaves are edible and sometimes used as vegetable.
- Vines are fed to animals.
- Antioxidants, which also serve as natural inflammatories are abundant in Sweet Potatoes. They also contain a large amount of Vitamin A and Vitamin C, which are also known to be natural inflammatory.

❖ **Tapioca**
  - Its fresh tubers are commercially consumed in Sago industries.
  - In addition, the boiled tubers are consumed as staple food largely on the West Coast.
  - Dried chips are also useful as cattle and poultry feed.
  - The fresh tubers and leaves contain harmful 'Hydrocynic acid' and hence are to be either sun-dried for 5-7 days or boiled for 15 minutes for human consumption or cattle purpose.

❖ **Elephant foot yam**
  - Its corms are used as vegetable and pickled for future use. Tender petioles are also edible.
  - Elephant foot is a rich source of vitamins A and B and minerals.
  - Culinary Purpose.
  - As medicine in many Ayurvedic preparation.
  - Recommended in case of piles, dysentery, asthma, swelling of lungs, vomiting, abdominal pain & as blood purifier.
  - To reduce pain in arthritis.
  - Animal feed. (leaves & stem)
  - Some species for decoration.
  - To add taste in 'Dal'.

❖ **TARO (COLOCASIA)**
  - Colocasia  is recommended for gastric patients and Colocasia flour is a good baby food.
  - In Hawaii and Polynesia, a fermented product prepared from Colocasia is very popular.
  - The pressure cooked Colocasia corms after being passed through strainer are allowed to ferment giving an acidic product called 'poi' in Africa.
  - The corm paste prepared from the cooked Colocasia is taken in the name 'fufu'.
  - In India it is used as a vegetable. Due to acridity, tubers are used as vegetable after thorough cooking.

➢ **Area, production and productivity of important underground vegetable crops**

| Sr. no | Crops | Area ('000' ha) | Production ('000' MT) | Productivity (tonne / ha) | Source |
|---|---|---|---|---|---|
| 1 | Onion | 1225 | 20991 | 17.13 | (NHB, 2015-16) |
| 2 | Garlic | 262.06 | 1425.46 | 5.44 | (NHRDF, 2015-16) |
| 3 | Radish | 193 | 2743 | 14.21 | (NHB, 2015-16) |
| 4 | Carrot | 79 | 1254 | 15.87 | (NHB, 2015-16) |
| 5 | Beetroot | 5 | 0.9 | 0.18 | (Textbook of vegetables, tubercrops and spices, S Thamburaj) |
| 6 | Turnip | 2.5 | 0.5 | 0.2 | (Textbook of vegetables, tubercrops and spices, S Thamburaj) |
| 7 | Potato | 2134 | 43770 | 20.51 | (NHB, 2015-16) |
| 8 | Sweet potato | 130 | 1472 | 11.32 | (NHB, 2015-16) |
| 9 | Cassava | 204 | 4554 | 22.32 | (NHB, 2015-16) |
| 10 | EFY | 426 | 8199 | 19.24 | (NHB, 2015-16) |

The cost of inorganic fertilizers has been enormously increased to an extent that they are out of reach of the poor, small and marginal farmers. It has become impractical to apply such costly inputs to the crop of marginal returns. These are biologically active products or microbial inoculants of bacteria, fungi, algae that may help either directly or indirectly in the enrichment of soil fertility. Biofertilizers are widely adopted as low cost supplements to chemical fertilizers and have no harmful effects either on soil health or environment. Research finding revealed that 25% of the N and P could be met through Biofertilizers for cultivated crops in our country.

Integrated Nutrient management practice of underground vegetables crops responds very well to organic manure. Therefore, the soil for underground vegetables crops like onion ,garlic, potato, sweet potato, tapioca, carrot, radish, turnip, yams etc. is liberally manures. Underground vegetables  being high demanding not adequately fertilizer, considerable yield losses are apparent the present day modern agriculture depend heavily on use of chemical fertilizers for boosting crop yield. However, indiscriminate use of fertilizers has an adverse effect on long term soil health and environment which has global attention.

The realistic solution is Integrated Nutrient Management system are the combined application of chemical fertilizers, longing with organic manure, green manure, bio-fertilizer and other organic recyclable materials for crop production. The basic concept underlying INM is the maintenance and adjustment of soil fertility and plant nutrients supply to an optimum level for sustainable. The desired crop productivity and soil health through optimization of the benefits from all possible sources of plant nutrients in an integrated manner (Roy and Ange, 1991).

Farmyard manure is conspicuous organic compost of an integrated nutrient supply system which improves soil health and releases macro and micro nutrient. FYM maintain soil fertility and water holding capacity. It also improves soil structure and texture. It also increases organic matter in the soil. Similarly, phosphate solubilizing bacteria play a significant role in solubilizing insoluble phosphate to make available around 95- 99% of the total soil phosphorus are insoluble and which are not directly available to plant. The phosphate solubilizing bacteria and fungi may convert insoluble form of phosphorus to soluble form by producing organic acid in general about 15-25% insoluble phosphate can be solubilized and 10-20% increase growth and yield production. Saving chemical fertilizers significantly, *Bacillus polymyxa*, *Aspergillus awamori*, *Penecillium digitatum*, are some important micro-organism (Bhattacharyya *et al.*, 2000).

Integrated nutrient management being moderate input to provide highly economic assurance, ecofriendly environment system soil health and plant growth by adding organic fertilizers like FYM, green manure and bio-fertilizer (*Azotobacter*, PSB). As well as supplementary of chemical fertilizer like N, P and K given by urea, DAP and murate of potash. It has produced sum growth promoting substances like Indole Acetic Acid (IAA), gibberellins, cytokinin, vitamins which help in root and shoot development and increase growth and yield production and productivity, enhance germination, flowering, maturation as well as better utilization of applied plant nutrient the growth period of crops to the bio-fertilizer bacteria secrete some fungicide and antibiotic substances which help in reducing occurrence of certain crops decreases and increases disease resistant in plants. These bacteria improve physical and biological properties, increase soil fertility and productivity, Organic fertilizers also improved water holding capacity.

**What is INM ?**

Integrated nutrient management is the maintenance or adjustment of soil fertility and plant nutrient supply at an optimum level to sustain the desired crop productivity.

This is done through optimization of the benefits from all possible sources of plant nutrients in an integrated manner.

In other words, integrated nutrient management is the use of different sources of plant nutrients integrated to check nutrient depletion and maintain soil health and crop productivity.

**Why is INM needed ?**

The increasing use of chemical fertilizers to increase the production of food and fibre is causing concern for the following reasons :

- Soils which receive plant nutrients only through chemical fertilizers are showing declining productivity despite being supplied with sufficient nutrients.
- The decline in productivity can be attributed to the appearance of deficiency in Secondary and micronutrients.
- The physical condition of the soil is deteriorated as a result of long-term use of chemical fertilizers, especially the nitrogenous ones.
- Excess nitrogen use leads to groundwater and environmental pollution apart from destroying the ozone layer through $N_2O$ production.

- The recent energy crisis, high fertilizer cost and low purchasing power of the farming community have made it necessary to rethink alternatives.
- Unlike chemical fertilizer, organic manure and biofertilizer available  locally at cheaper rates. They enhance crop yield per unit of applied  nutrients by providing a better physical, chemical and microbial environment. This ultimately improves crop yield.
- The available quantity of animal excreta and crop residues cannot meet the country's requirements for crop production. Therefore, maximizing  the usage of organic waste and combining it with chemical fertilizers and Biofertilizers in the form of integrated manure appears to be the best alternative.
- 

**Different components of INM**

- There are various components of plant nutrients for INM which can be applied in an integrated way. Besides inorganic fertilizers as the major component, others include farmyard.
- Manure (FYM), composts, green manure crops, crop residues, crop rotation and bio fertilizers. Fertilization in a balanced way, improved crop nutrition maintain the soil fertility and of plant nutrient supply to an optimum level for sustaining the desired crop productivity through optimization of various plant nutrients in an integrated manner.

i. Chemical fertilizers: Chemical fertilizers are rich in nutrients. They are required in less quantity to supply nutrients as compared to organic manures. But continuous use of chemical fertilizers deteriorates the soil conditions. Therefore, chemical fertilizers should be accompanied by organic Biofertilizers.

ii. Organic manures like FYM in situ, Vermicompost: It improves the bulk density of soil up to a layer of 25 cm. It reduces resistance to penetration and Supplements N up to 50% of the nitrogenous requirement of the crop. Increases available N and P use efficiency when combined with 100% of the recommended quantity of NPK and Biofertilizers.

iii. Industrial waste various practices can be adopted to convert wastes into suitable products Convert all available biomass on the farm into compost instead of burning or wasting it.

iv. Inclusion of legume crops in cropping system to fix the atmospheric nitrogen in the soil.

v. Use of Biofertilizers like *azolla, blue green algae,* and *rhizobium* etc.

vi. Crop residues and make use of cattle excreta as manure rather than as fuel.

vii. Green manuring either growing in the same field or incorporating of leguminous plant or leaves.

viii. Crop rotations :It is most important INM strategy which is ignored by the growers that is crop rotation is a very important tool in sustaining nutrient supply. Legumes in rotation restore soil fertility in more than one way viz, some of the N fixed is left in the soil after harvest, improvement in soil properties, lesser disease and pest problem and better weed control.

**Importance of Integrated Nutrient Management**

- There is an urgent need to adopt an integrated nutrient supply and management system for promoting efficient and balanced use of plant nutrients.
- While the main emphasis was given on increasing the proper and balanced used of mineral fertilizers, the role of organic manure, Biofertilizers, green manuring and recycling of organic wastes should be considered supplementary and not substitutable.
- On the one hand, there is a vast scope for increasing plant nutrient supply through the use of organic fertilizers, but there is, on the other hand, no scope for reducing the consumption of mineral fertilizers since the present level of crop productivity has to be increased in the coming years.
- Owing to the growing demand for more agricultural yields/products and the scarcity of land resources, attention is placed more on intensification of farming systems in the country.
- Intensification means a more educated group of farmers to be trained in good agricultural practices including appropriate Integrated Nutrient Management technologies.
- The national research institutes, including universities in close cooperation with the Department of Agricultural extension, private enterprise and NGO's have to play a vital role in the promotion of INM practices to farmers.
- The farmers are to be educated to practically think the nutrient providing capacity of organic manures, crop residues, composts and Biofertilizers.
- Farmers have to be trained for efficient use of locally available organic manures and Biofertilizers most suitable to the needs of the area and the cropping system as a whole.
- But now we are to think to educate farmers to make the use of organic, inorganic and biological fertilizers.
- Plant nutrient in future will require judicious and integrated management of all sources of nutrients.
- It is, therefore, necessary to solve the problems of plant nutrition in an integrated way and to maintain the overall balance and flow of soil nutrients.
- There is an urgent need to adopt an integrated nutrient supply and management system for promoting efficient and balanced use of plant nutrients.
- While the main emphasis was given on increasing the proper and balanced used of mineral fertilizers, the role of organic manure, Biofertilizers, green manuring and recycling of organic wastes should be considered supplementary and not substitutable.
- On the one hand, there is a vast scope for increasing plant nutrient supply through the use of organic fertilizers, but there is, on the other hand, no scope for reducing the consumption of mineral fertilizers since the present level of crop productivity has to be increased in the coming years.
- This will not be possible without the use of mineral fertilizer, as long as no other practical low input technology has  become available for higher productivity.
- Effective implementation of such improved agricultural practices, using a combine approach requires skilled management and adaptiveness by farmers who need to  be educated and trained, assisted by national research and extension services.

- The time has come to effectively coordinate the efforts of all agencies/institutions to benefit agriculture.
- 

**Concept of INM**

India is predominantly an agriculture-based country and more than two-third of the population depends on agriculture for their livelihood. India with geographical area of 329 M ha presently supports 17% of the world's population on merely 2.5% world's land area and 4% world's fresh water resources.

India made a spectacular achievement in attaining the self sufficiency in food production by the introduction of high yielding dwarf and fertilizer responsive varieties of cereals, particularly wheat and rice in the mid- 1960s .

With the use of improved varieties coupled with increased fertilizer and agro-chemicals use, price support and other policy initiatives, the food grain production increased from 50.8 mt in 1951 to 213.18 mt during 2003 – 04.

Despite this impressive achievement in food grain production the per unit productivity of most of the crops is still very low as compared to other countries.

In the early 1990s, however, fertilizer became the target of criticism, mainly because of heavy use in the developed countries, where it was suspected of having an adverse effect on the environment through nitrate leaching, eutrophication, greenhouse gas emissions and heavy metal uptakes by plants.

Consequently, fertilizer use per se was mistakenly identified as harmful to the environment. But, if for any reason fertilizer use were discontinued today, world food output would drop by an estimated amount of 40 per cent.

While fertilizer misuse can contribute to environmental contamination, it is often an indispensable source of the nutrients required for plant growth and food production.

Unless all the soil nutrients removed with the harvested crops are replaced in proper amounts from both organic and sustained; soil fertility will decline.

If in the past, the emphasis was on increased use of fertilizer; the current approach should aim on educating farmers to optimize use of organic, inorganic and biological fertilizer in an integrated way.

Plant nutrition to day requires judicious and integrated management of all sources of nutrients for sustainable agriculture.

**How INM differs from conventional farming?**

Integrated nutrient management differs from conventional nutrient management in that it considers nutrients from different sources, not only organic materials, nutrients carried over from previous cropping seasons, transformation of nutrients in soil, In conventional farming, people gave more emphasis on grain yield through use of chemical fertilizers, use of high yielding varieties and chemical pesticides along with irrigation facilities.

In INM it integrates/combines the objectives of production with ecology and environment, that is, optimum crop nutrition, optimum functioning of the soil health, and minimum nutrient losses or other adverse effect on the environment.

Integrated Nutrient Management (INM) has to be considered an integral part of any sustainable agricultural system.

**Principles of INM**

The basic principle underlying INM in the maintenance and possible increase of soil fertility for sustaining increased crop productivity through the use of all possible sources, organic and inorganic, of plant nutrients required for crop growth and quality in all integrated manner appropriate to each cropping system and farming situation within the given ecological, social and economic boundaries.

Attempts have been in our country to complement the use of mineral with organic sources of plant nutrients generated useful, though information on the complementary and synergistic effects of these materials on the yield of crops. Because organic sources of nitrogen are also improving soil structure and soil bioactivity which are not directly improved by mineral sources of N. The productivity of the crop for each kg of N may be better with organic sources than sources of N.

If the objective of INM is the balanced and effective use of various sources of plant nutrients then the strategy should be the mobilization of all available, accessible and affordable plant nutrient sources in order to optimize the environmentally safe productivity of the whole cropping system and to increase the monetary return to the farmer.

Thus, there is need for information on:

I. Integrated nutrient recommendations for cropping systems as a whole taking into account the complementary and the synergistic effects of combined use of both mineral and organic/biological sources for sustained crop production.
II. Recommendations for different agro-ecological situations taking into account available organic/biological resources.
III. Transfer of this technology for the benefit of small farmers through the national agricultural extension services.

**Future strategies:**
- Emphasis should be given to INM research in horticultural crops.
- Direct and indirect benefits of INM on a long term basis needs to be quantified.
- More effective ways of converting organic wastes into manures are to be evolved.
- Promotion of bio-fertilizers.
- Demonstrations for spreading the concept and technologies of INM.

**REVIEW**

Jamir *et al.* (2013) at Nagaland conducted an experiment on effect of integrated nutrient management on growth, yield and quality of onion (*Allium cepa* L.).) and they reported that significantly maximum plant height (55.38 cm), number of leaves (9.80), neck thickness (1.52 cm) were recorded with treatment $T_7$ (50 % NPK + 50 % FYM), while bolting per cent found non significant.

Jamir *et al.* (2013) at Nagaland university, Nagaland conducted an experiment on effect of integrated nutrient management on growth, yield and quality of onion (*Allium cepa* L.) and they reported that significantly maximum bulb diameter (5.58 cm), weight of bulb

(64.40 g) yield (18.06) and quality perameter significantly maximum TSS (13.18$^0$ brix) were recorded with treatment $T_7$ (50 % NPK + 50 % FYM).

Jamir *et al.* (2013) at Nagaland university, Nagaland conducted an experiment on effect of integrated nutrient management on growth, yield and quality of onion (*Allium cepa* L.) and they reported that highest yield (18.06 t/ ha), gross income (180670 Rs/ ha), net income (129260 Rs/ ha) and cost benefit ratio (1:3.5) were recorded with treatment $T_7$ (50 % NPK + 50 % FYM).

Shahi (2013) at Varanasi conducted an experiment on effect of integrated use of organic manures, inorganic fertilizers and bio-fertilizers on growth and yield of onion and they reported that significantly maximum plant height at harvesting (51.90 cm), bulb weight (177.0 g), bulb diameter (7.87 cm) and yield of onion (74.85 q / ha) were recorded with treatment of $T_6$ (50% N through vermicompost + 25% N through urea + PSB + *Azotobacter*).

GARLIC

Bhandari *et al.* (2012) at Anand conducted an experiment on effect of integrated nutrient management on growth and quality of garlic and they reported that significantly maximum plant height at 50 DAP (40.27 cm), number of leaves/plant at 50 DAP (6.82), diameter of stem at 50 DAP (1.14 cm), diameter of bulb (5.34 cm) and length of bulb (4.50 cm) were recorded with treatment $T_9$ (100-40-60 NPK kg/ha + 100 kg N/ha through C. C. + *Azotobacter* + PSB).

Bhandari *et al.* (2012) at Anand conducted an experiment on effect of integrated nutrient management on yield and quality of garlic and they reported that significantly maximum number of cloves /bulb (19.33), average weight of bulb (36.94 g), yield per hectare (124.56 q), TSS content (45.05 °Brix) and sulphur content of bulb (0.94 %) were recorded with treatment $T_9$ (100-40-60 NPK kg/ha + 100 kg N/ha through C. C. + *Azotobacter* + PSB).

Nainwal *et al.* (2015) at Lucknow (U.P) conducted an experiment on response of garlic on integrated nutrient management practices in a sodic soil of Uttar Pradesh and they reported that significantly maximum fruit weight (52.43 g), number of cloves per plant (37.66), dry matter (53.24 g) and yield (60.86 q ha$^{-1}$) were recorded with treatment $T_6$ (100 % NPK +FYM @ 20 t ha$^{-1}$ + PSB + *Trichoderma*).

RADISH

Mohammad *et al.* (2015) at Etawah ( U. P.) conducted an experiment on effect of integrated nutrient management on growth and yield attributes of radish ( *Raphanus sativus* L.) and they reported that significantly highest yield (843.43 q/ha) were recorded with the treatment $T_8$ [75 % NPK (80:60:60 kg/ ha) + FYM (10 t/ ha) + PSB (5 kg /ha)].

CARROT

Vithwel *et al.* (2013) at Nagaland  conducted on experiment on integrated nutrient management on productivity of carrot and fertility of soil and  they reported that significantly maximum plant height (25 cm), root length (18.88 cm), root diameter (4.14 cm), root weight (90.37 g), root yield (30.08 t/ha) and carotene content (3.41 mg $100^{-1}$g) were recorded with treatment of  $T_9$ (50% NPK + 50% FYM + Biofertilizers).

Anand *et al.* (2016) at Bangalkot (Karnataka) conducted an experiment on integrated nutrient management in carrot (*Daucus carota*) under north estern transitional track of Karnataka and  they reported that significantly maximum plant height (30.40 cm), root growth (15.42 cm), root girth (10.40 cm), root yield (24.10 t/ha) were recorded with treatment $T_7$ (50% RDF + 25% N through FYM + 25% N through vermicompost).

TURNIP

Yanthan *et al.* (2012) at Nagaland conducted an experiment on effect of integrated nutrient management on growth, yield and nutrient uptake by turnip (*Brassica rapa* L.) and  they reported that significantly maximum plant height (50.16 cm), number of leaves (14.43) were recorded with treatment of $T_8$ (50 % Pig manure + 50 %  NPK) , while significantly maximum leaf area (214.20 $cm^2$) was recorded with treatment T9 (50 % Poultry manure + 50 %  NPK).

Yanthan *et al.* (2012) at Nagaland conducted an experiment on effect of integrated nutrient management on growth, yield and nutrient uptake by turnip (*Brassica rapa* L.) and they reported that significantly maximum fresh weight of root (245.12 g), length of roots (16.24 cm), diameter of root (8.43 cm), root yield (522.51 q $ha^{-1}$) and TSS (4.32 $^{o}$Brix) were recorded with treatment of $T_8$ (50 % Pig manure + 50 %  NPK).

POTATO

Yadav *et al.* (2014) at CPRI, Shilong  conducted an experiment on effect of integrated nutrient management on production of seed tubers from true potato (*Solanum tuberosum*) they reported that significantly maximum plant height (38.4 cm), shoots per plant (1.7), leaf area index (3.2),tuber per plant (5.8), yield per plant (71.1 g) and tuber weight (12.4 g), number of tubers (984.2 $\times 10^3$/ ha) and tuber yield (12.40 t/ha)  were recorded with treatment of $T_2$ (75% RDF and 25% N through FYM).

Banerjee *et al.* (2016) at West Bengal conducted an experiment on effect of integrated nutrient management in potato based cropping system in alluvial soil  of West Bengal and they reported that significantly maximum germination per cent (93.37 %), yield (25.80 t $ha^{-1}$), net return (28650 $ha^{-1}$) and B:C ratio (1.23) was recorded with  $O_2F_2$ [FYM 20 t $ha^{-1}$ + 100% RDF, (200:150:150 NPK kg $ha^{-1}$)] while significantly maximum Total no. of tubers (463734 $ha^{-1}$) was recorded with $O_3F_2$ [Residue incorporation + bio-fertilizers + 100% RDF(200:150:150 NPK kg $ha^{-1}$)].

SWEET POTATO

Allolli *et al.* (2011) at Dharwade conducted an exdpriment on effect of integrated nutrient management (INM) on yield and economics of sweet potato (*Ipomoea batatas* L.) and they reported that significantly superior with respective to yield per plot and heacter (24.16 kg and 33.55 tones ha$^{-1}$) were recorded with treatment of $T_{10}$ (RD FYM 10 t ha$^{-1}$+ 50:25:50 Kg NPK ha$^{-1}$)

**EFY**

Sarvaiya *et al.*(2010) at Navsari conducted an experiment on Influence of integrated nutrient management (INM) on growth and yield parameters of elephant foot yam under south Gujarat condition and they reported that significantly highest corm yield (55.33 t / ha) was recorded with treatment of $T_7$ (100% RDF (Through IOS) + *Azospirillum* 5 kg ha-1 + PSB 5 kg ha-1) while all growth parameter found non significant.

GREATER YAM

Kaswala *et al.* (2013) at Navsari conducted an experiment on organic production of greater yam yield, quality, nutrient uptake and soil fertility and they reported that significantly maximum tuber yield (16.4 t/ha), vine yield (34.72 kg/ha) and B:C ratio (2.7) were recorded with treatment $T_9$ (FYM @ 10 t/ha + NPK @ 80:60:80 kg/ha) while, starch (14.1 %), carbohydrate (15.2%) recorded with $T_1$ (Biocompost (50%) + castor cake (50%) (biocompost @ 3.38 t/ha + castor cake @ 1.35 t/ha)

**CASSAVA**

Ashok *et al.* (2013) at Andhra Pradesh conducted an experiment on integrated nutrient management for Cassava under rainfed condition of Andhra Pradesh and they reported that significantly maximum plant height (372.7 cm), number of tubers (13.0 plant$^{-1}$), tuber girth (21.0 cm), tuber yield (33.6 t / ha) were observed with treatment of $T_6$ (Sunhemp @ 50 kg ha$^{-1}$ + RD of K + 50% RD of NP + *Azospirillum* + *phosphorus solubilising bacteria* (PSB) @ 5 kg ha$^{-1}$ each) while significantly maximum stem girth (10.7 cm) and starch (26.5 %) were recorded with $T_7$ (Dhaincha @ 50 kg ha$^{-1}$ + RD of K + 50% RD of NP+ *Azospirillum* + PSB @ 5 kg ha$^{-1}$ each).

Mhaskar *et al.* (2013) at Dapoli (Maharashtra) conducted an experiment on integrated nutrient management for sustainable production of cassava in Konkan region and they reported that significantly maximum plant height (183.11 cm), number of main branches (All the treatment combination produced on an average tow main branches per plant ), leaf length (27.89 cm) were recorded with $T_8$ (¾ RD of FYM + NK + vermicompost @ 200 kg ha$^{-1}$) while significantly maximum number of sub branches (12.00) were recorded with treatment $T_4$ (¾ RD of FYM + NK + green leaf manure *Glyricidia* @ 25 t ha-1) + 3 % Panchagavya).

Mhaskar *et al.* (2013) at Dapoli (Maharashtra) conducted an experiment on integrated nutrient management for sustainable production of cassava in Konkan region and they reported that significantly maximum tuber length (29.98 cm), tuber yield (2.91 kg/plant) and weight of tuber (441.41 g ) were recorded with $T_4$ (¾ RD of FYM + NK + green leaf manure (*Glyricidia* @ 25 t ha$^{-1}$) + 3 % Panchagavya) while significantly maximum tuber girth (17.64 cm) were recorded with treatment of $T_3$ (¾ RD of FYM + NK + dhaincha @ 20 kg ha$^{-1}$(green manure) + 3 % Panchagavya).


**TURMERIC**

S.P. Singh (2012) at Muzaffarpur conducted an experiment on effect of integrated nutrient management on growth, yield and economics of turmeric(Curcuma longa L) var.RAJENDFRA SONIA and they reported that significantly maximum plant height (121.18 cm), no. of leaves per tiller (10.28), yield(54.93 t/ha) and cost benefit ratio (1:4.46) was recorded with treatment $T_2$ [200 q ha-1 FYM +1/2 N,P,K (RDF) + P-solublizers, *Pseudomonas f luorescence* and *Trichoderma* as seed treatment and soil application @ 20 kg ha-1 each bio-fertilizers with spray or drenching with mancozeb (Indofil M-45) @ 0.25 per cent and malathion @ 0.1 per cent at 21 days interval from July-October for controlling disease and insect pest].

**GINGER**

Singh *et al.* (2015) at Katni (M.P.) conducted an experiment on effect of integrated nutrient management on growth, yield and economics of ginger and they reported that significantly maximum weight of tuber (229.9 plant$^{-1}$), yield (22754 kg / ha) and value cost ratio (4.12) were recorded with treatment $T_5$ (Improved cultivar *Suprabha*, NPK @ 75 : 50 : 50 kg/ha and FYM @ 20 t/ha in partial shade).

**Conclusion**

From the foregoing discussion, it can be concluded that different level of fertilizers, organic manures and biofertilizers either alone or in combination significantly increased the growth, yield and quality of the underground vegetable crops.

| Sr. No | Crop | Recommended Treatments | Result |
|---|---|---|---|
| 1. | Onion | 50 % NPK + 50 % FYM(20 t ha$^{-1}$) (120:60:60 NPK kg ha$^{-1}$) | Increase bulb weight, bulb diameter and yield (18.06 t ha$^{-1}$). |
| 2. | Garlic | 100-40-60 NPK kg/ha + 100 kg N/ha through C. C. + *Azotobacter* + PSB (*Azotobacter* and PSB were applied 5 ml/kg seed ) | Increase the plant height, marketable yield (124.56 q/ha) and TSS. |
| 3. | Radish | 75 % NPK (80:60:60 kg/ ha) + FYM (10 t PSB(5kg /ha) | Increase root yield (843.43 q/ha). |
| 4. | Carrot | 50 % RDF (NPK @ 80:60:60 kg/ha) + 50 % FYM (10 t /ha) + Rhizosphere bacteria | Increase plant height, root growth, and yield (30.08 t /ha ). |
| 5. | Turnip | 50 % Pig manure (20 t ha$^{-1}$) +50 % NPK(80:40:40 kg ha$^{-1}$) | Increase plant height ,fresh weight of roots, root length , diameter of roots and yield (522.51 q/ ha). |
| 6. | Potato | FYM 20 t ha$^{-1}$ + 100% RDF (200:150:150 NPK kg ha$^{-1}$) | Increase plant height , shoots per plant, ,number of tubers per plant and yield (25.80 q/ ha). |
| 7. | Sweet potato | FYM 20 t ha$^{-1}$ | Increase yield (31.99 t/ha) |
| 8. | Elephant foot yam | 50% RDF (Through inorganic source) + 50% RDN (Through organic source: FYM) + *Azospirillum* 5 kg ha$^{-1}$ | Increase yield (42.77 t/ ha). |
| 9. | Greater yam | FYM @ 10 t/ha + NPK @ 80:60:80 kg/ha | Increase tuber yield and vine yield (16.4 t/ha). |
| 10. | Cassava | ¾ RD of FYM + NK + vermicompost @ 200 kg ha$^{-1}$ + 3% Panchagavya | Increase plant height, number of tubers, tuber girth and yield (45.67 t/ha). |
| 11. | Turmeric | 200 q ha$^{-1}$ FYM +1/2 N,P,K (RDF) + P-solublizers, *Pseudomonas fluorescence* | Increase height of plant, number of tiller, number of |

| | | and *Trichoderma* as seed treatment and soil application @ 20 kg ha$^{-1}$ each bio-fertilizers with spray or drenching with mancozeb (Indofil M-45) @ 0.25 per cent and malathion @ 0.1 per cent at 21 days interval from July-October for controlling disease and insect pest. | leaves and yield (54.93 t/ha). |
|---|---|---|---|
| 12. | Ginger | Improved cultivar Suprabha, NPK @ 75 : 50 : 50 kg/ha and FYM @ 20 t/ha in partial shade. | Increase weight of rhizome and yield (22754 kg/ha). |

**Future strategies:**
- Emphasis should be given to INM research in horticultural crops.
- Direct and indirect benefits of INM on a long term basis needs to be quantified.
- More effective ways of converting organic wastes into manures are to be evolved.
- Promotion of bio-fertilizers.
- Demonstrations for spreading the concept and technologies of INM.

**References :**

Allolli, T.B., Athani, S.I. and Imamsaheb, S.J. (2011). Effect of integrated nutrient management (INM) on yield and economics of sweet potato (*Ipomoea batatas* L.). *The Asian Journal of Horticulture,* Vol. **6** (1): 218-220.

Annonymous (2016). *Indian Horticulture Databse,* National Horticulture Board, Department of Agriculture and Cooperation, Government of India.

Annonymous (2016). *The National Horticultural Research and Development Foundation.* Department of Agriculture and Cooperation, Government of India.

Ashok, P., Ramanandan, G., Sasikala, K. (2013). Integrated nutrient management for Cassava under rainfed condition of Andhra Pradesh. *Journal of Root Crops,* Vol. **39** (1):100-102.

Banerjee, H., Sarkar, S., Ray, K., Rana, L. and Chakraborty, A. (2016). Effect of integrated nutrient management on emergence, total number of tubers and yield of tubers (cv. Kufri Jyoti) and net returns of potato. *Annuals of plant and soil research,* Vol. **18** (1) : 8-13.

Bhandari, S.A., Patel, K.S., and Nehete, D.S. (2012). Effect of integrated nutrient management on yield and quality of garlic. *The Asian Journal of Horticulture,* Vol. **7** : 48-51.

Bhattacharyya, P. and Kumar, R. (2000). Liquid biofertilizer current knowledge and future prospect. *National seminar on development and use of biofertilizer, biopesticides and organic manures.* West Bengal 10-12.

Jamir, S., Singh, V.B., Kanaujia, S.P. and Singh, A.K. (2013). Effect of integrated nutrient management on growth, yield and quality of onion (*Allium cepa* L.). *Progressive Horticulture,* Vol. **45** (2).

Kaswala, A.R., Kolambe, B.N., Patel, K.G., Patel, V.S. and Patel, S. Y. (2013). Organic production of greater yam: yield, quality, nutrient uptake and soil fertility. *Journal of root crops,* Vol. **39** (1): 56-61.

Khalid, M., Yadav, B.K. and Yadav, M.P. (2015). Effect of integrated nutrient management on growth and yield attributes of radish ( *Raphanus sativus* L.). *Annals of horticulture,* Vol. **8** (1): 81-83.

Mhaskar, N.V., Jadye, A.T., Haldankar, P.M., Chavan, S.A. and Mahadkar, U.V. (2013). Integrated nutrient management for sustainable production of cassava in Konkan region. *Journal of Root Crops,* Vol. **39** (1) : 42-47.

Nainwal, R.C., Singh, D., Katiyar, R.S., Sharma, L., and Tewari, S.K. (2015). Response of garlic on integrated nutrient management practices in a sodic soil of Uttar Pradesh, India. *Journal of Spices and Aromatic Crops,* Vol. **24** (1) : 33-36.

Patil, A.G., Mangesh and Rajkumar, M. (2016). Integrated nutrient management in carrot (*Daucus carrota* L.) under north estern transitional track of Karnataka. *An International Quarterly Journal of life sciences,* Vol **11** (1) : 271-273.

Roy, R.N., and Ange, A.L. (1991). Integrated plant nutrition system and sustainable agriculture. *FAI Annual Seminar,* pp S V/ 1-12.

Saravaiya, S.N., Chaudhary, P.P., Patel, D.A., Patel, N.B., Ahir, M.P. and Patel, V.I. (2010). Influence of integrated nutrient management (INM) on growth and yield parameters of elephant foot yam under south Gujarat condition. *The Asian Journal of Horticulture,* Vol **5** (1): 58-60.

Shahi, S.K. (2013). Effect of organic manures, inorganic fertilizers and biofertilizers addition on soil properties and productivity under onion ( *Allium cepa* L.). *Plant Archives,* Vol. **13** (1) pp. 381-387.

Singh, A.K., Gautam US., Singh, J. (2015). Effect of integrated nutrient management on growth, yield and economics of ginger. *Bangladesh journal of botany,* **44** (2) : 341-344.

Singh. S.P. (2012). Effect of integrated nutrient management on growth, yield and economics of turmeric. *The Asian journal of horticulture,* Vol. (**7**): 478-480.

Thamburaj, S., Narendrasingh, N. (2014). Textbook of Vegetables, Tuber crops and Spices Crops. Directorate of knowledge management in agriculture. 156-161.

Vithvel and Kanaujia, S.P. (2013). Integrated nutrient management on productivity of carrot and fertility of soil. *SAARC Journal of Agriculture.,* **11** (2): 173-181.

Yadav, S.K., Srivastava, A.K., and Bag, T.K. (2014). Effect of integrated nutrient management on production of seed tubers from true potato (*Solanum tuberosum*) seed. *Indian Journal of Agronomy* **59** (4): 646-650.

Yanthan, T.S., Singh, V.B., Kanaujia, S.P. and Singh, A.K. (2012). Effect of integrated nutrient management on growth, yield and nutrient uptake by turnip (*Brassica rapa* L.) cv. Pusa sweti and their economics. *Journal of soils and crops,* Vol **22** (1): 1-9.